I am STEM
Vol. 2: African-American Men in STEM

By: Corell Oglesby

Copyright © 2025 by Corell Oglesby
ISBN: 979-8-218-79927-4

Blessed Publications

I am a Scientist
Bernard Harris Jr.

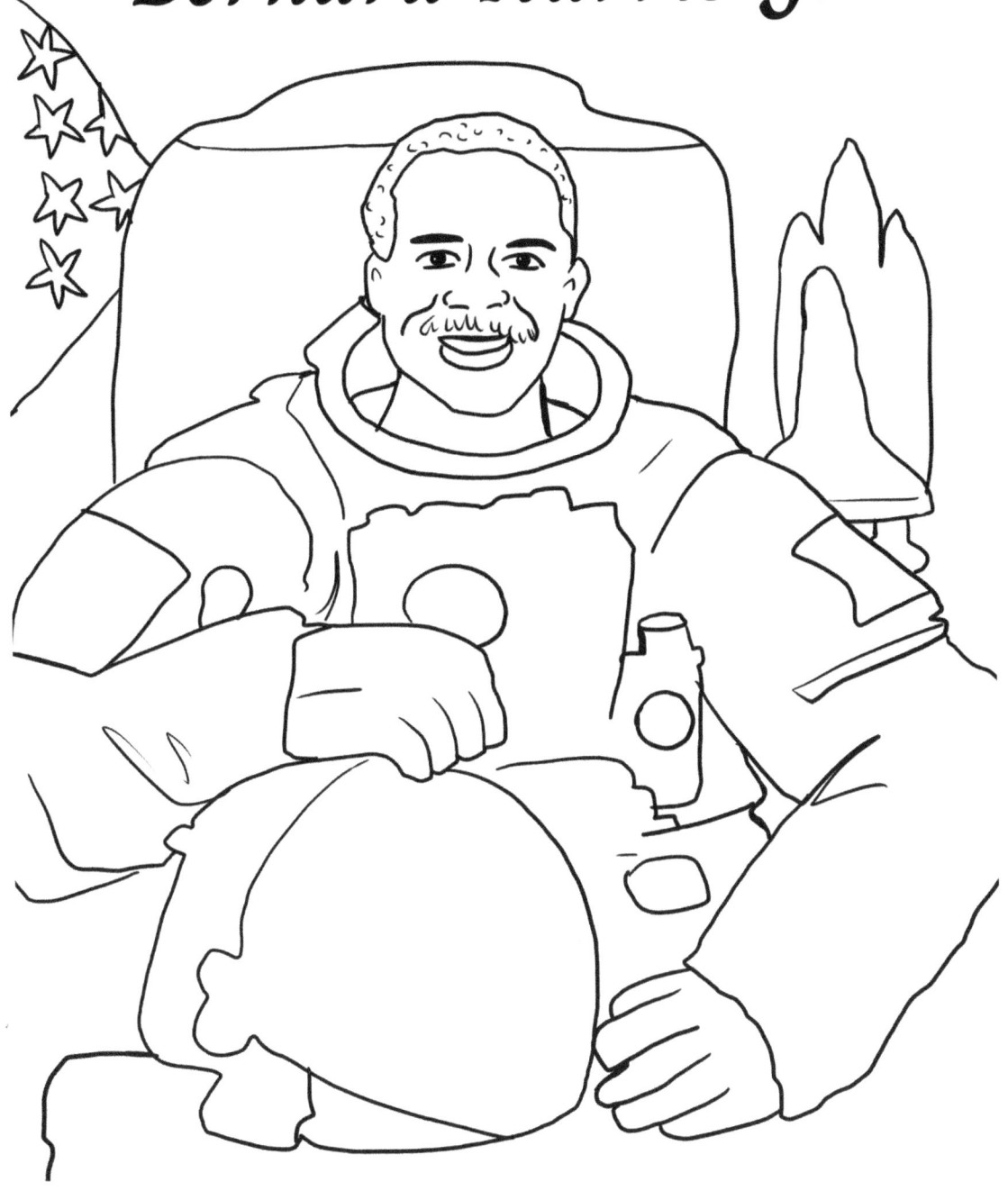

Bernard Anthony Harris Jr. was an astronaut for NASA. In 1995, during his second Space Shuttle mission, he made history by performing a spacewalk and becoming the first African American to do so.

3

Charles Drew

Charles Richard Drew helped create and implement large-scale blood banks, which proved essential in saving thousands of Allied forces' lives during World War II.

Daniel Hale Williams

Daniel Williams established the first non-segregated hospital in the United States and founded a nursing school for African-Americans. As a physician, he was also one of the pioneers in performing successful open-heart surgery.

Edward Bouchet

Edward Bouchet made history in 1876 as the first African-American to receive a Ph.D. from a university in the United States. Bouchet shared his knowledge as a chemistry and physics teacher for over two decades.

6

Ernest Everett Just

Ernest Just was the first African American to earn a Ph.D. from the University of Chicago in 1916 and is known for his groundbreaking research in the field of cell biology along with his contributions to the understanding of fertilization and embryonic development.

George Washington Carver

Carver was an accomplished American agricultural scientist and inventor who advocated for alternative crops to cotton and soil preservation techniques. His most notable achievement was his groundbreaking work with peanuts, earning him the nickname "Peanut Man." Carver's research led to the development of over 300 peanut-based products.

Jesse Ernest Wilkins Jr.

Jesse Wilkins Jr. was a writer of scientific papers, textbooks, and an accomplished mathematician. He was also a part of the Manhattan Project, which developed the first atomic bomb, although he wasn't aware of this fact until after the bomb was deployed. He served as a professor of mechanical engineering at Howard University for many years.

Norbert Rillieux

Norbert Rillieux was a chemical engineer whose ideas revolutionized the sugar refining process. His patented technique not only made the refining process safer for workers, but it also greatly improved efficiency. In addition to his contributions to the sugar industry, Rillieux developed a plan to reduce mosquito breeding grounds in response to the yellow fever outbreak in New Orleans.

Percy Lavon Julian

Percy Julian's accomplishments in medicine include creating artificial human hormones. In addition to his achievements in the laboratory, Julian was a strong advocate for diversity and inclusion within the scientific community.

Samuel Massie Jr.

Samuel Proctor Massie, Jr. was a notable chemist. After graduating from high school at age 13, he went on to work on the renowned Manhattan Project. Massie was also the first African American to teach at the esteemed United States Naval Academy.

12

I am a Technologist
Clarence Ellis

Clarence Ellis made history when he became the first African-American to earn a Ph.D. in computer science from the University of Illinois. He created the first Groupware systems. These collaborative software tools and applications were designed to facilitate communication and collaboration among groups and individuals working together on shared projects.

Mark Dean

Mark Dean's educational background includes graduating from the University of Tennessee and earning his Ph.D. from Stanford. He is best known to have over 25 patents credited to his name. One of his most notable accomplishments includes co-creating the IBM personal computer.

Thomas Mensah

Thomas O. Mensah was a distinguished inventor and chemical engineer who has made significant contributions to the development of fiber optic manufacturing and nanotechnology.

I am an Engineer
Alexander Miles

Alexander Miles invented the automatic elevator door. He received a patent for his invention in 1887, and his design has been the basis for almost all automatic elevator doors since then.

Elijah McCoy

Elijah McCoy moved to the United States from Canada in 1847. He obtained a certification as a mechanical engineer. McCoy went on to file more than 50 patents, including one for his groundbreaking invention of the automated lubrication system for steam engines.

Frederick McKinley Jones

Born in Kentucky in 1893, Frederick Jones became a prolific inventor with over 60 patents to his name. His most famous invention was the mobile refrigeration system for trucks, which he developed while co-founding the company Thermo King.

James West

With over 250 patents to his name, James West is credited with designing the modern-day microphone, among other things. In addition to his many achievements, West is also passionate about promoting diversity in STEM fields, and has created several programs to support this important cause.

Garrett Morgan

Born in 1877 in Kentucky, Garrett Morgan secured his first patent in 1912. In 1914, he created the first gas mask which was originally called a "smoke hood." Morgan continued to innovate and patent a number of other products, including the traffic light.

20

Lewis Howard Latimer

Latimer was an inventor who made significant contributions to the toilet, the telephone and the light bulb. In addition to this work, Latimer also made improvements to the design of railroad cars.

Lonnie G. Johnson

Lonnie Johnson is an accomplished engineer and inventor. He attended the prestigious Tuskegee University and worked at NASA before creating his most famous invention, the Super Soaker. In 1995, he sold the rights to his invention to Hasbro.

Otis Boykin

Otis Boykin earned an electrical engineering degree from Fisk University in Nashville, Tennessee. Boykin's inventions included a control unit for pacemakers, a variable resistor for guided missile systems, and a burglar-proof cash register.

Paul E. Williams

The first functional helicopter is said to have been invented by Paul E. Williams, who was born in Birmingham, Alabama, in 1939. Williams' helicopter was a small aircraft that was able to lift itself off the ground and hover in the air for several seconds.

I am a Mathematician
Benjamin Banneker

Born in 1806 in Maryland, Banneker was a noteworthy astronomer, mathematician, and author. He is recognized for his outstanding contributions to surveying and predicting solar eclipses, as well as for authoring the almanac series.

David Blackwell

David Blackwell made significant contributions to the fields of probability theory, statistics, and game theory. He became the first African American to be inducted into the National Academy of Sciences and served as a professor at prestigious institutions.

Elbert F. Cox

Cox was born in 1895 in Indiana. Despite being a talented violinist and receiving a scholarship to attend the Prague Conservatory of Music, Cox pursued a career in Mathematics. In 1925, he received his Ph.D. in Mathematics from Cornell University, becoming the first African American to do so.

Keywords

Vocabulary

1. STEM: An acronym for Science, Technology, Engineering, and Mathematics, representing a group of academic disciplines that involve problem-solving, innovation, and critical thinking.

2. Pioneers: Individuals who lead the way by exploring new territories, challenging conventions, and making groundbreaking discoveries or inventions.

3. Innovation: The process of creating new ideas, products, or methods that improve upon existing ones, often involving creative thinking and problem-solving.

4. Curiosity: A strong desire to learn, explore, and understand the world, often by asking questions and seeking answers.

5. Creativity: The ability to think imaginatively, generate original ideas, and approach challenges from unique perspectives.

6. Scientist: A person who conducts research, experiments, and investigations to gain a better understanding of the natural world and its phenomena.

Vocabulary

7. Chemist: A scientist who specializes in the study of the composition, properties, and reactions of substances and the transformation of matter.

8. Engineer: A professional who applies scientific principles and mathematics to design, build, and develop solutions for practical problems and innovations.

9. Inventor: A person who creates new devices, products, or systems through a combination of creative thinking and practical knowledge.

10. Impact: The effect or influence that an action, discovery, or invention has on individuals, communities, or society as a whole.

11. Dream: A personal aspiration or ambition that motivates and guides an individual toward a specific goal or vision.

12. Vision: A clear and inspiring mental image of a future achievement or outcome, often serving as a guiding force for one's actions and efforts.

13. Curriculum: A structured plan or course of study, often used in educational settings to guide learning and development in specific subjects.

Vocabulary

14. Resource: A tool, reference, or material that provides information, support, or assistance in various activities, including learning and research.

15. Glossary: A list of important terms and their definitions, often included at the end of a book or document to assist readers in understanding key concepts.

16. Diversity: The presence and inclusion of a variety of different backgrounds, cultures, and perspectives, often leading to greater creativity and innovation in STEM fields.

17. Inclusivity: The practice of ensuring that individuals from all backgrounds, regardless of gender, race, or other factors, have equal opportunities and representation in STEM.

18. Mentorship: The guidance, support, and advice provided by experienced individuals (mentors) to help others (mentees) in their personal and professional development.

19. Exploration: The act of seeking new knowledge or experiences by venturing into unfamiliar territory, whether it's scientific exploration or discovering new ideas.

Vocabulary

20. Achievement: The successful completion or accomplishment of a goal, task, or objective, often resulting from hard work, dedication, and perseverance.

Match the Word

Match the Word

Instructions:

1. Match each word in the left column with its correct definition by writing the letter of the definition next to the corresponding word.

2. Discuss your answers with a partner to see if you both agree on the definitions!

3. Reflect on how these concepts are important in STEM and in your everyday life.

Match the Word

1. **Curiosity**

2. **STEM**

3. **Pioneers**

4. **Innovate**

5. **Creativity**

6. **Impact**

7. **Invent**

8. **Dream**

9. **Vision**

10. **Explore**

A. The ability to generate new and original ideas or solutions.

B. To create or design something that has not existed before.

C. A strong desire to learn or know something

D. The act of investigating something new

E. Individuals who are among the first to explore or innovate in a particular field.

F. An acronym for Science, Technology, Engineering, and Mathematics, focusing on problem-solving.

G. The effect or influence that something has on a situation, person, or community.

H. A desired goal.

I. The ability to think about or plan the future with imagination or wisdom.

J. To travel through an unfamiliar area to learn about it.

Puzzles

Word Search

Puzzle #1

```
I M P A C T C O A I P C I R I
E X P E R I M E N T I R O Y N
T C V K T W Q N S S O T B B C
J O K F C W O E T B N V S N L
U D O K W V W U O E E J G O U
V I E L A C N T V Y E Y Y I S
I N J T S A I N G S R N T T I
J G I S E C I O O F S Z I A V
O O D P S I L V Y F E E V R I
N Y T I S O I R U C A C I O T
E N G I N E E R I N G N T L Y
M A N H G O A L S B T E A P B
S A C E W W O L H N M I E X O
G E T S R E T U P M O C R E H
T W W H M B V R U S M S C W M
```

coding	computers	creativity
curiosity	engineering	experiment
exploration	goals	impact
inclusivity	innovation	inventor
math	peanuts	pioneers
robotics	science	technology
tools		

Word Search

Puzzle #2

```
K L P A A M M E F F S R P L P
E T I R W K S E T M R E M E S
A N M M I R L X D C A E V N U
U R K R U R O A E I O N C G P
A S T N E T A P W U C O R I E
M O Y Q R K U Q O E T I B N R
E J S E D O C T O R C P N E S
U K A J E D N X Y H C A N E O
A D O T N O J F E I X G P R A
R O X M H O C M F L I Q B S K
T H C P S H I F H N W U Z I E
A N F G Y S A S E Y D J I L R
W N J R T R T S I N I L O I V
H M K G T P U G V V Z T D O B
P G E K S H X M A J L I G H T
```

art	chemist	doctor
engine	engineer	hood
light	medicine	nurse
patents	pioneer	read
smoke	spacewalk	supersoaker
traffic	violinist	vision
write		

38

Math Problems

Math Problems

1. $\dfrac{2}{2} - \dfrac{1}{2} =$ _____

2. $\dfrac{1}{3} + \dfrac{1}{3} =$ _____

3. $\dfrac{5}{7} - \dfrac{1}{7} =$ _____

4. $\dfrac{1}{2} + \dfrac{1}{2} =$ _____

5. $\dfrac{8}{9} - \dfrac{3}{9} =$ _____

Math Problems

6. $\frac{1}{2} \times \frac{1}{2} =$ _____

7. $\frac{1}{2} \times \frac{1}{3} =$ _____

8. $\frac{1}{5} / \frac{1}{10} =$ _____

9. $\frac{1}{2} / \frac{1}{2} =$ _____

10. $\frac{3}{4} \times \frac{2}{5} =$ _____

Science Experiment

Science Experiement

Objective: To encourage readers to become plant scientists, observe plant growth, and understand the importance of nurturing the Earth. The majority of these items can be found at your local dollar store. Note: A parent may need to assist with this assignment.

Materials Needed:
1. Small potted plants (one per child or group)
2. Soil
3. Watering can or spray bottle
4. Sunlight or a well-lit area
5. Notebook and writing utensil

1. Introduction:

Welcome, scientists! Today, we're embarking on a fantastic journey into the world of plants and becoming plant scientists just like the famous George Washington Carver. Have you ever wondered how plants grow, why they're so important, or how they can change the world? Well, you're about to find out!

Plants are amazing living things that play a crucial role on our planet.

Science Experiement

They give us oxygen to breathe, provide us with food to eat, and make our world beautiful with their vibrant colors and shapes. But have you ever thought about how they do all of this?

George Washington Carver was a scientist who had a special love for plants. He discovered incredible things about them that helped farmers and improved people's lives. He showed us that plants are not just green things in the ground; they are living beings with their own secrets.

Today, you're going to become plant scientists. We'll give each of you a small potted plant, and together, we'll learn how to take care of it, observe it, and help it grow. Just like George Carver, we'll be curious, ask questions, and discover the wonders of the plant world.

Are you ready to become plant investigators? Let's grab our magnifying glasses and notebooks, put on our science caps, and start this exciting adventure into the world of plants. Remember, every great scientist starts by being curious, and today, you'll take your first step into the amazing world of plant science!

Science Experiement

2. Selecting Plants: Select the type of plant you would like to grow. These can be houseplants or outdoor plants, depending on the location and season. You can do research on the types of plants that can be grown in different planting zones.

3. Observation: Let's do some observation. Examples of questions to answer are below. Feel free to make other observations not listed below.

What do you notice about the leaves?
Are there any flowers or buds?
Is the soil dry or moist?
How much sunlight does your plant receive?

4. Recording Observations: Use your notebook or journal and draw or write about your observations. You can create a plant diary where you record their findings over several days or weeks.

5. Watering: Make sure you water your plant according to the instructions you found during your research.

Science Experiement

6. Sunlight: Discuss the role of sunlight in plant growth. Ensure that the plants are placed in a well-lit area or near a window to receive the necessary sunlight according to your research.

7. Growth Monitoring: Over a period of time (days or weeks), regularly check and record the changes in your plants. You can measure the height of the plant, count the number of leaves, or note any new growth.

8. Discussion: If you would like, discuss your observations with others. Example questions are listed below.

What changes have you noticed in your plant?
How does watering and sunlight affect plant growth?
Why is it important to take care of plants?

9. Reflection: Reflect on what you have learned about plant growth and the importance of caring for the Earth, just like George Washington Carver did with his agricultural innovations. Parents may want to assist with this activity.

Let's Code with Scratch

Let's Code with Scratch

Welcome to Scratch, a fun and creative platform for coding and making your own interactive stories, games, and animations! Whether you're a kid or just a kid at heart, Scratch is a fantastic way to learn programming concepts and express your creativity. Here's how to get started:

Step 1: Access Scratch

- Open your web browser and go to the Scratch website: https://scratch.mit.edu.

- Create an account with Scratch by clicking the Join Scratch button. If you already have an account, sign in.

- Click on the Create button to start a new project.

Step 2: To understand the Scratch

- Go to Interface go to https://en.scratch-wiki.info/wiki/Scratch_Wiki:Table_of_Contents/Program to learn more about Scratch.

- Stage: This is where your projects come to life. Characters and objects (sprites) interact on the stage.

- Sprites: These are the characters, objects, or elements you can control and animate. You can add, delete, or create sprites.

Let's Code with Scratch

- <u>Blocks Palette:</u> This is where you'll find all the coding blocks you can use to create scripts.

- <u>Coding Area:</u> Drag and snap coding blocks together to create scripts. This is where you'll write your code.

- <u>Costumes and Backdrops:</u> Customize your sprites and stage with different costumes and backdrops.

- <u>Sound Library:</u> Add sound effects and music to your projects from the sound library.

Step 3: Creating Your First Script

Now, let's create a simple script to get you started:

- Select a sprite by clicking on it in the Sprites list.

- In the Blocks Palette, choose a block from the Events category, like when green flag clicked.

- Add more blocks, like move 10 steps from the Motion category, under the when green flag clicked block.

- Click the green flag icon in the stage area to run your script. Your sprite should move when you click the green flag!

49

Let's Code with Scratch

Step 4: Experiment and Explore

- Experiment with different blocks to create animations, games, and stories.

- You can change a sprite's appearance, move it, make it speak, and more!

- Explore the other categories like Looks, Sound, and Control to discover more possibilities.

Step 5: Save and Share Your Project

- Click the File menu, then choose Save now to save your project.

- To share your project, click the Share button and follow the instructions to make it public or share a link.

Step 6: Learn and Have Fun

- Scratch offers a wealth of resources. Explore the Tips section and the Scratch Wiki for guidance.

- Try out other Scratch projects created by the community for inspiration and learning.

Remember, coding is all about experimenting, making mistakes, and finding creative solutions. Have fun, and let your imagination run wild with Scratch!

Let's Code with Scratch

Use the open box after the Edit option to rename the project. Also, click the Share button to allow others to see your project.

Costume   Sounds

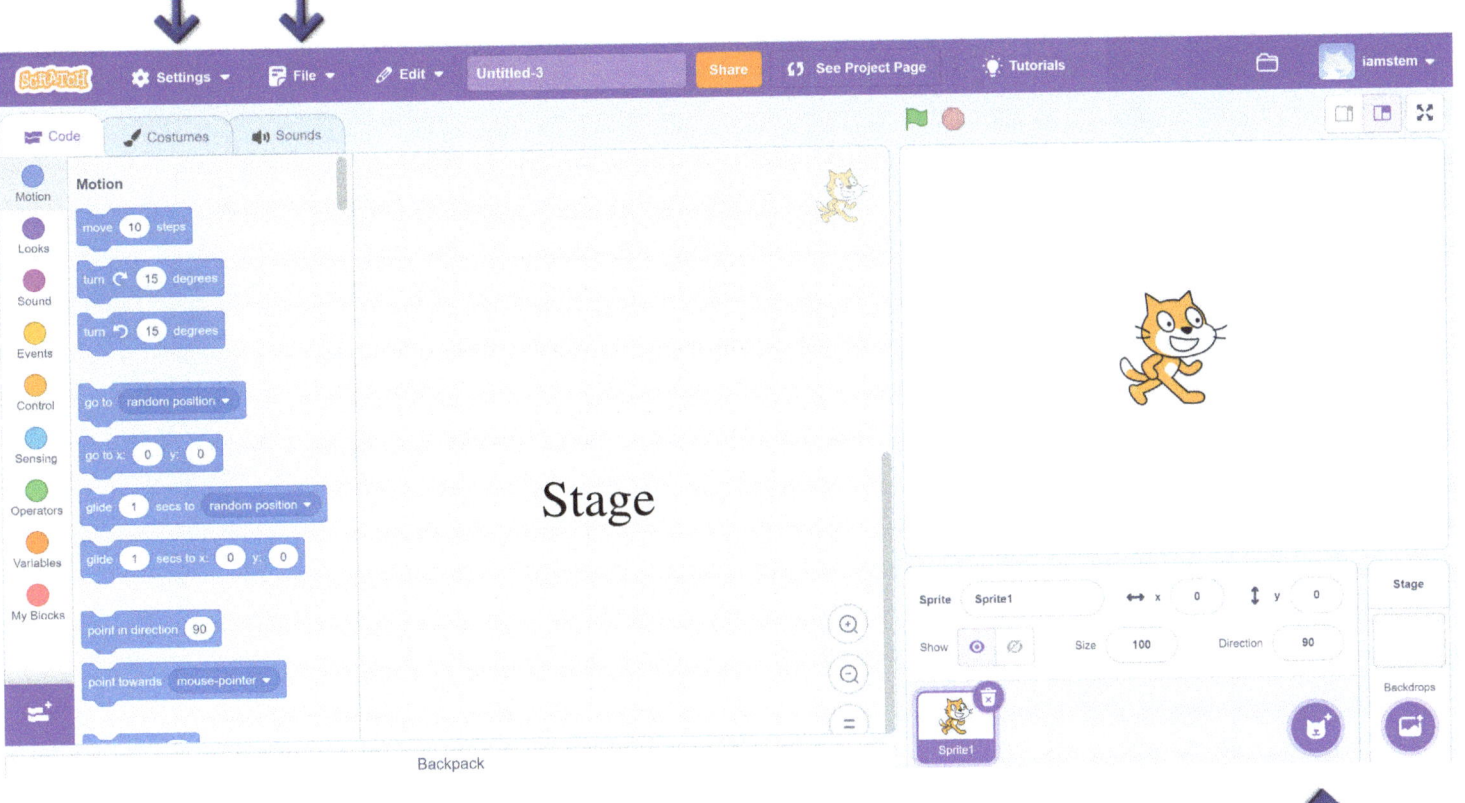

Stage

Block Palette

Sprite

Backdrop

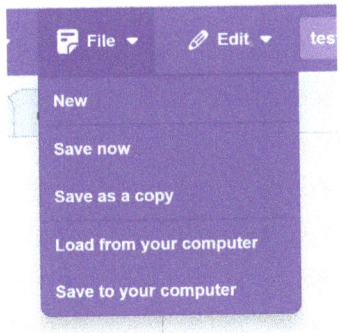

Use the File option to create a new file, save the current file or save as a copy. Save to or load from a computer is an option as well.

51

Let's Code with Scratch

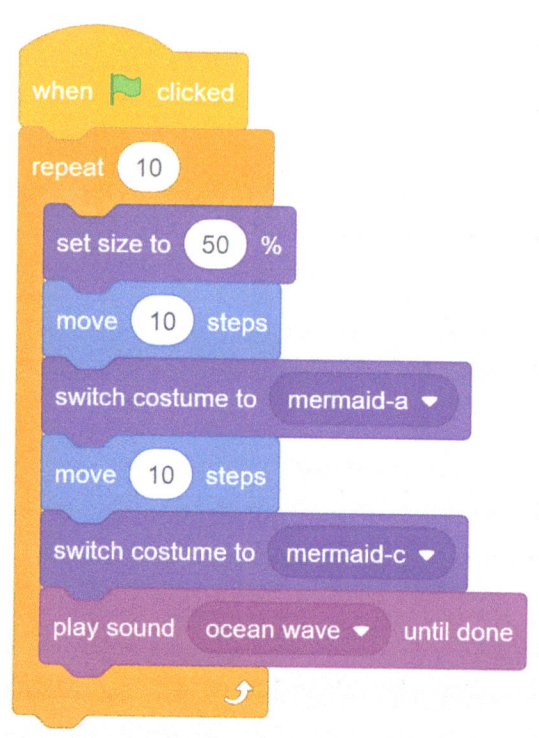

Example code is provided. We chose a mermaid sprite and underwater 1 as the background. This code shows the mermaid gliding through the ocean. Scan the QR code to go to the project below.

52

Let's Code with Scratch

Let's create a Game. This game is called Guess the Number!

- Be sure to rename this project from Untitled to Guess the Number or to a name that you prefer.
- Choose a sprite from the sprite library.
- Choose a backdrop.
- Click on Events category and drag when green flag clicked block to the scripting area.
- Create the following variables: Guess and Number.
  - This can be found in the orange variables area.
  - Click the button: Make a Variable.
  - Make this option available for all sprites.
- Set the Number to Pick a random number between 1 to 50.
- Say Pick a random number between 1 to 50. You have 10 tries. Set this for 5 seconds. To make a sprite speak, go to the Look category.
- Go to Control category and choose repeat. Update the number to 10 times.
  - The following instructions will be nested inside the repeat block.

Let's Code with Scratch

- In the Sensing category, choose the ask and wait block. Type in the empty box What is your number?
- In the variables section, choose the set option.
- Using the drop down, select guess.
- Set guess to answer. Answer is found in the Sensing category.
- You will need 3 If then blocks. These blocks can be found in the Control category.
- If the answer = number, then you will say You are correct! for 5 seconds.
- At the top of the page is whele the different sound options are found. Find and choose the Clapping sound.
- Go to the Sound area in the coding block area. Choose play sound. Click the drop down and choose clapping.
- If the answer < number, then you will say Incorrect! for 5 seconds.
- At the top of the page is where the different sound options are found. Find and choose the Bonk sound.
- Go to the Sound area in the coding block area. Choose play sound. Click the drop down and choose bonk.
- If the answer > number, then you will say Incorrect! for 5 seconds.
- At the top of the page is where the different sound options are found. Find and choose the Bonk sound.

Let's Code with Scratch

- Go to the Sound area in the coding block area. Choose play sound. Click the drop down and choose bonk.
- Then choose Stop All under the Control category, and place outside the Repeat block.
- Test your code by pushing the green Go flag at the top of the screen to ensure the code is working correctly.
- Go to the Solutions section to compare your code or for a helpful hint.
- Make alterations. Examples are below:
- Change the number from 50 to a higher number.
- Add a timer and guess the correct number within a certain time.
- Decrease or increase the guess count.
- Continue to test until there are no other changes that need to be made.

Invention Time

Create your own Game or Toy

Objective:

To foster creativity, innovation, and problem-solving skills in participants while encouraging them to explore the principles of design and engineering.

Materials Needed:

1. Paper and pencils for brainstorming and sketching
2. Craft materials such as cardboard, construction paper, markers, glue, scissors, tape, and any other crafting supplies you have available
3. 3D modeling software or clay for creating prototypes (Optional)
4. Access to a computer or smartphone for research and inspiration

Instructions:

1. Brainstorm Ideas:

- Begin by brainstorming ideas for a new toy or game. Think about what interests you and what types of toys or games you enjoy playing with.

Create your own Game or Toy

- Consider the age group and interests of the target audience for your toy or game. Is it designed for young children, teenagers, or adults?
- Write down any ideas that come to mind, no matter how wild or unconventional they may seem. This is the creative brainstorming stage, so let your imagination run wild!

2. Research and Inspiration:

- Take some time to research existing toys and games to see what's already out there. Look for inspiration in books, magazines, websites, and social media platforms.
- Pay attention to what makes certain toys or games popular and successful. Are there any common themes or features that stand out to you?
- Consider how you can put your own unique spin on a classic toy or game, or come up with something entirely new and innovative.

3. Design and Prototyping:

- Once you have a clear idea of what you want to create, start sketching out your design on paper. Think about the overall look and feel of your toy or game, as well as any specific features or mechanics it may have.

58

Create your own Game or Toy

- Use craft materials such as cardboard, construction paper, and markers to create a prototype of your toy or game. This could be a rough mock-up or a more detailed model, depending on your design and crafting skills.
- If you're comfortable with technology, you can also use 3D modeling software or clay to create a digital or physical prototype of your design.

4. Testing and Refinement:

- Once you have a prototype of your toy or game, it's time to test it out! Invite friends, family members, or classmates to try out your creation and provide feedback.
- Pay attention to how people interact with your toy or game. Are they having fun? Are there any issues or challenges they encounter while playing?
- Use this feedback to make any necessary improvements or refinements to your design. This might involve tweaking the rules of the game, adjusting the size or shape of the toy, or adding new features to enhance the gameplay experience.

Create your own Game or Toy

5. Finalizing Your Design:

- Once you're happy with your toy or game, it's time to finalize your design. Create a polished version of your prototype using your chosen materials and crafting techniques.

- Consider adding any finishing touches or decorations to make your toy or game stand out and appeal to your target audience.

- Take photos or videos of your finished creation to document the design process and showcase your invention to others.

6. Share Your Invention:

- Finally, share your invention with others! You can demonstrate your toy or game to friends, family members, or classmates, or even showcase it at a school or community event.

- Consider creating a presentation or video to explain the inspiration behind your invention, how it works, and why you think it's special.

- Who knows – your invention could be the next big hit in the world of toys and games!

Solutions

Math Problems

1. $\dfrac{2}{2} - \dfrac{1}{2} = \dfrac{1}{2}$

2. $\dfrac{1}{3} + \dfrac{1}{3} = \dfrac{2}{3}$

3. $\dfrac{5}{7} - \dfrac{1}{7} = \dfrac{4}{7}$

4. $\dfrac{1}{2} + \dfrac{1}{2} = \dfrac{2}{2}$ or 1

5. $\dfrac{8}{9} - \dfrac{3}{9} = \dfrac{5}{9}$

Math Problems

6. $\dfrac{1}{2}$ x $\dfrac{1}{2}$ = $\dfrac{1}{4}$

7. $\dfrac{1}{2}$ x $\dfrac{1}{3}$ = $\dfrac{1}{6}$

8. $\dfrac{1}{5}$ / $\dfrac{1}{10}$ = 2

9. $\dfrac{1}{2}$ / $\dfrac{1}{2}$ = 1

10. $\dfrac{3}{4}$ x $\dfrac{2}{5}$ = $\dfrac{6}{20}$ or $\dfrac{3}{10}$

Match the Word

1. **Curiosity - C**

2. **STEM - F**

3. **Pioneers - E**

4. **Innovation - D**

5. **Creativity - A**

6. **Impact - G**

7. **Invent - B**

8. **Dream - H**

9. **Vision - I**

10. **Explore - J**

Word Search

Puzzle #1

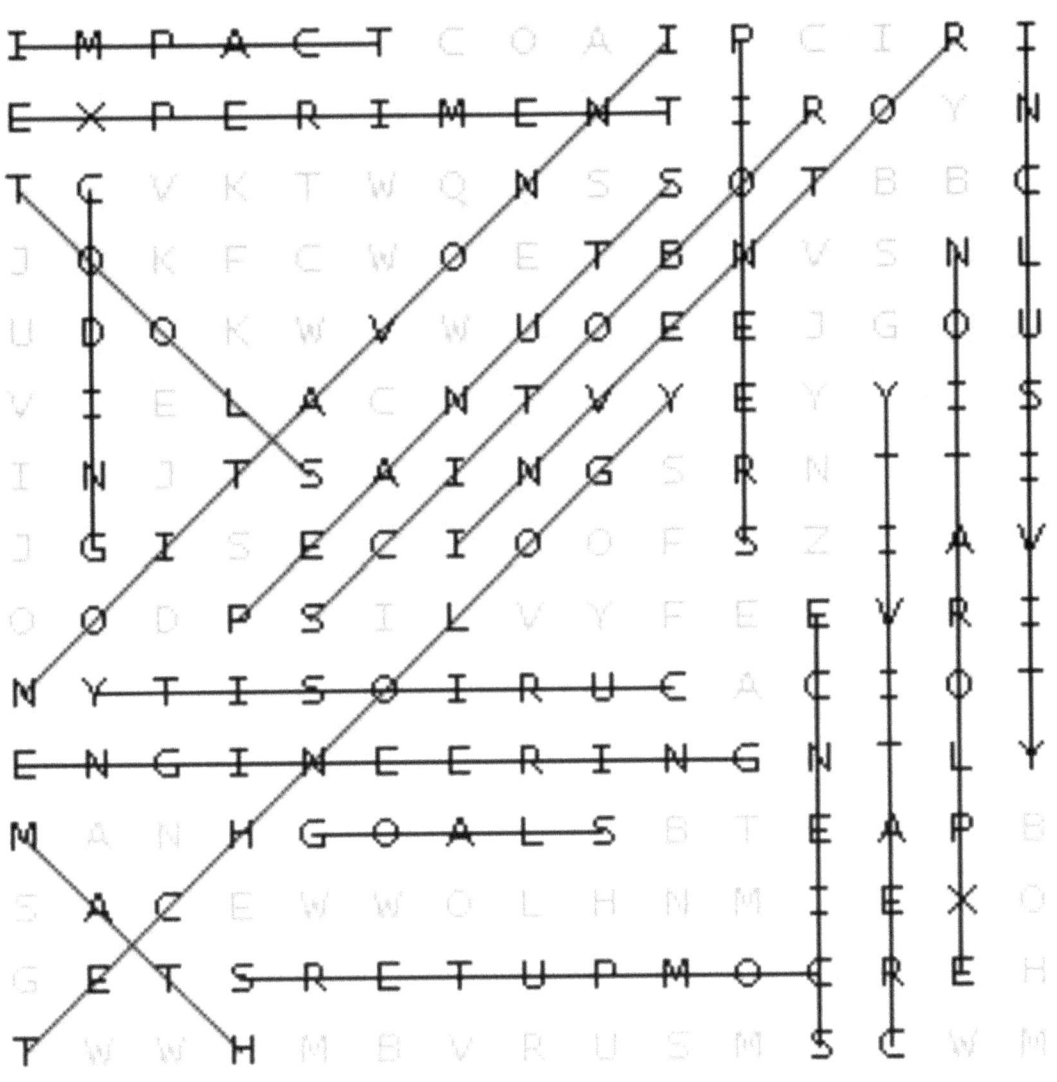

Word Search

Puzzle #2

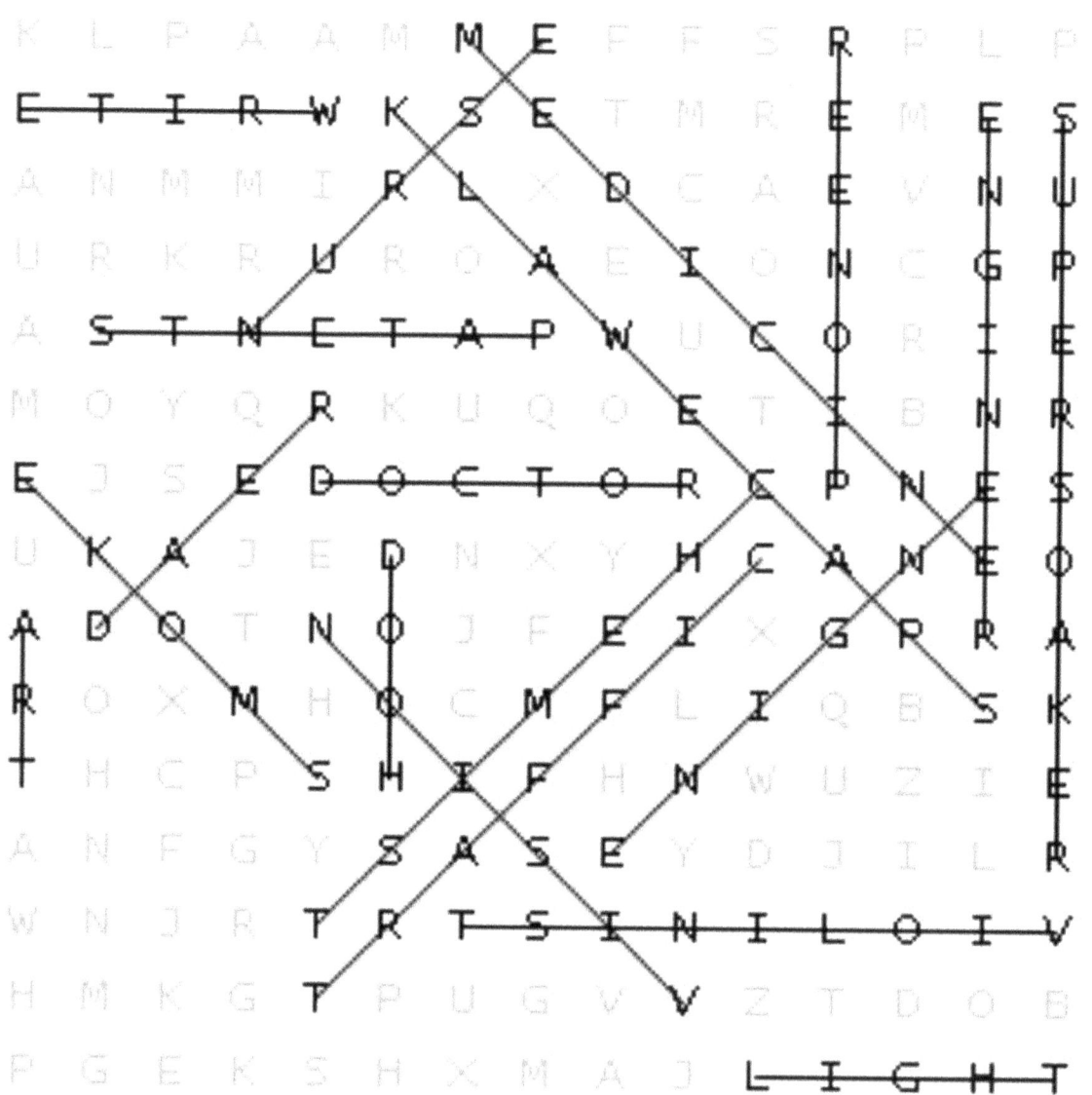

Let's Code with Scratch
Solutions

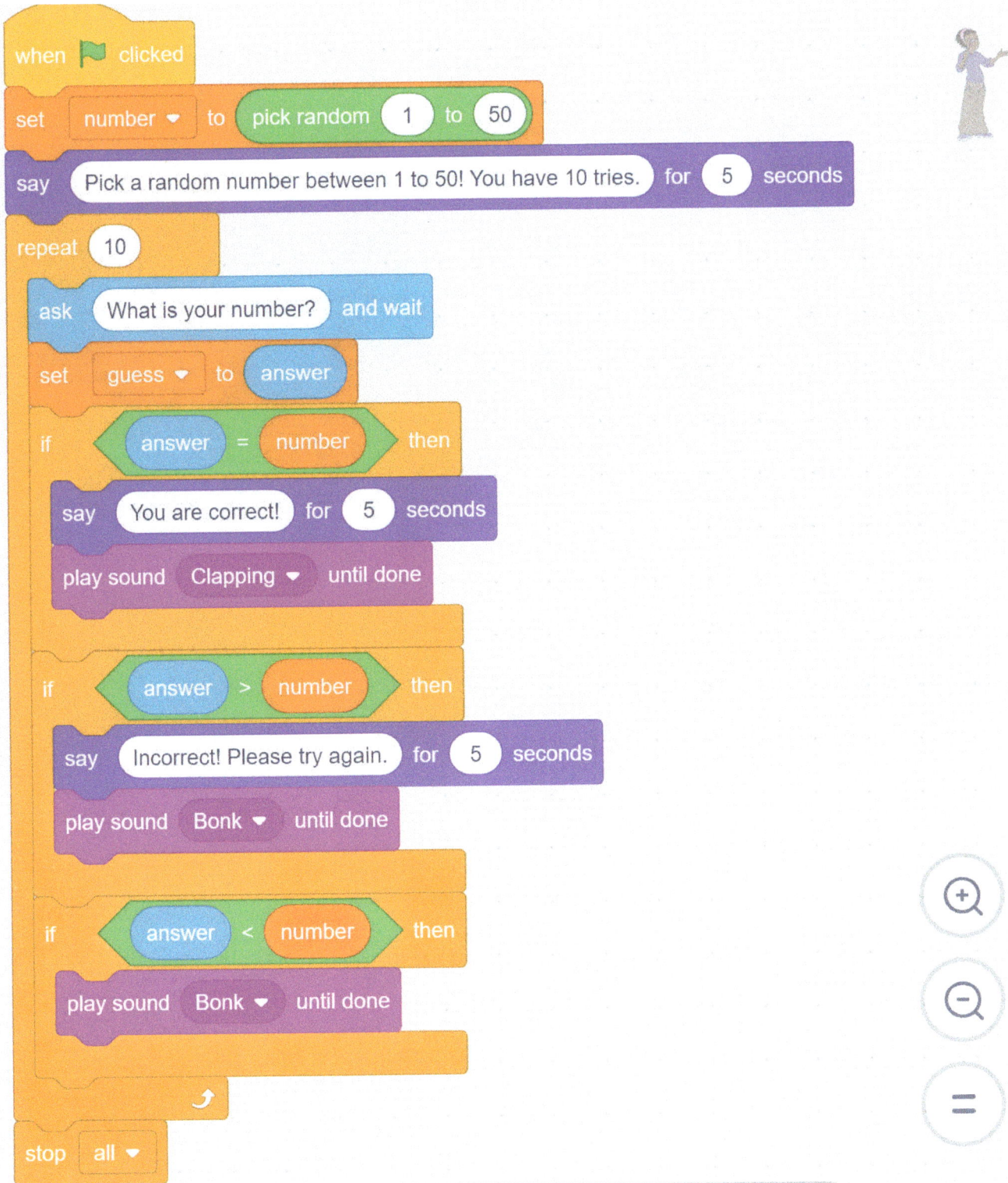

Let's Code with Scratch Solutions

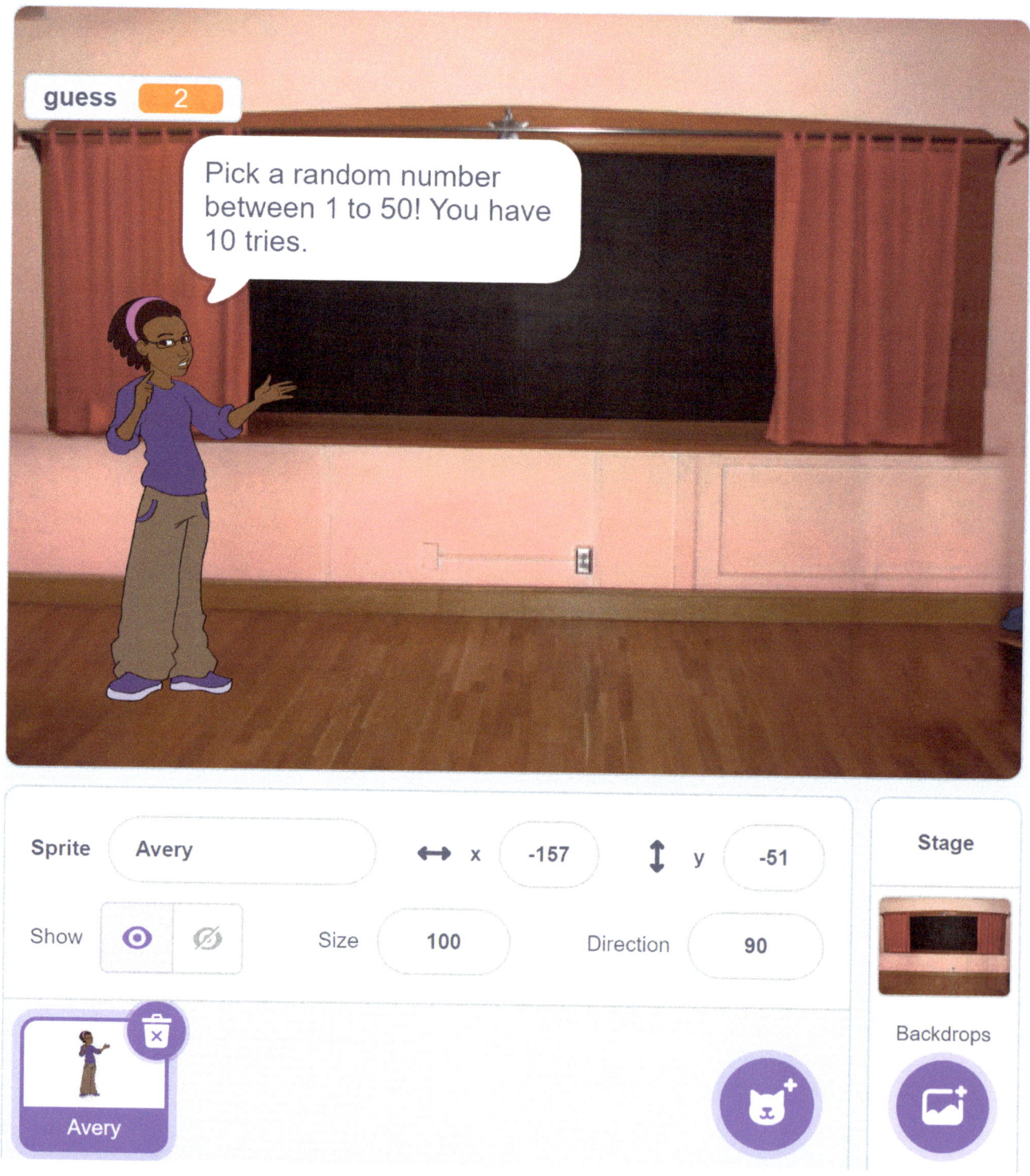

Let's Code with Scratch
Solutions

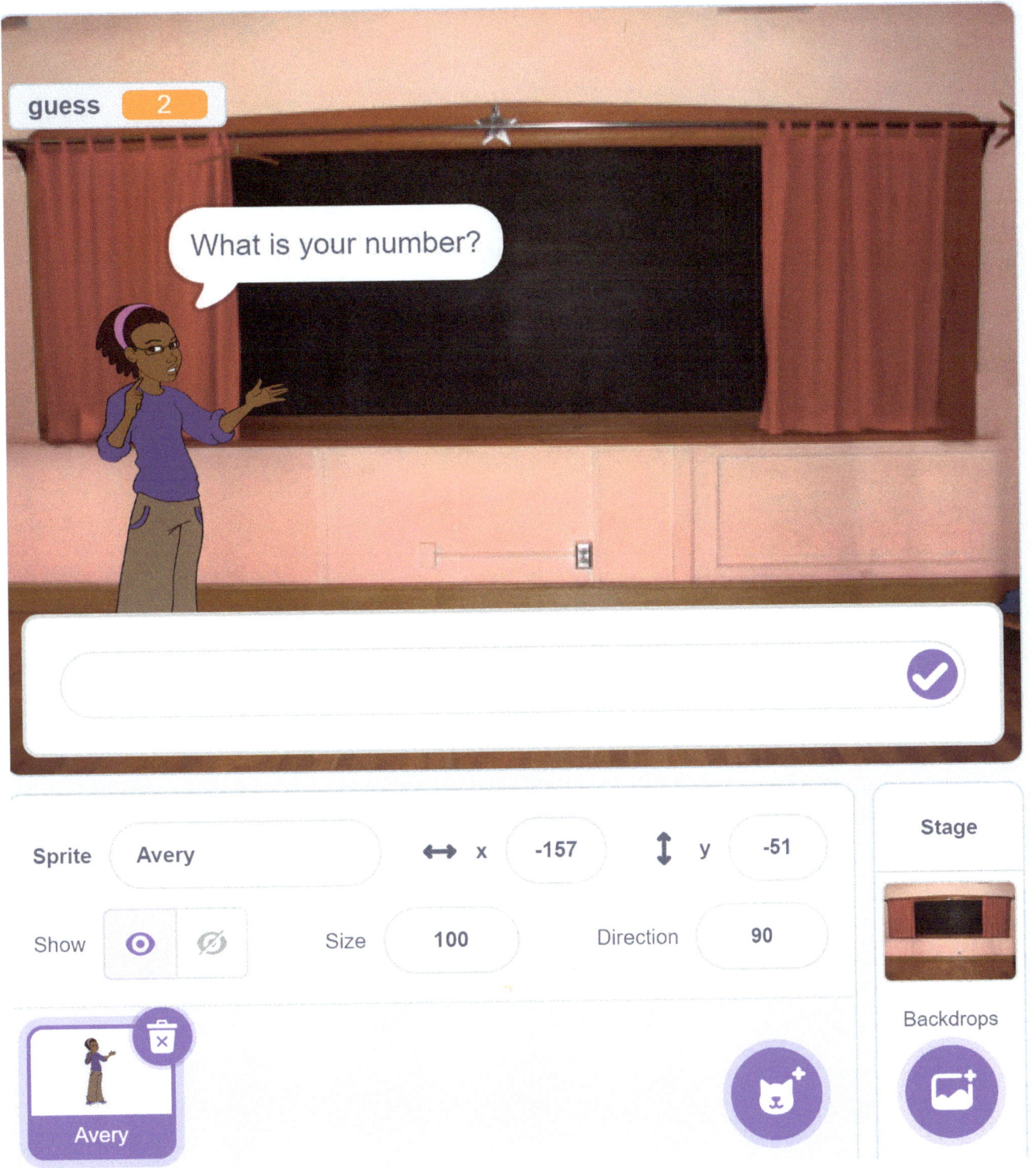

The End

www.ingramcontent.com/pod-product-compliance
Lightning Source LLC
Chambersburg PA
CBHW041124120626

46547CB00019B/2844